Editor
Lorin Klistoff, M.A.

Managing Editor
Karen Goldfluss, M.S. Ed.

Editor-in-Chief
Sharon Coan, M.S. Ed.

Cover Artist
Barb Lorseyedi

Art Coordinator
Kevin Barnes

Art Director
CJae Froshay

Imaging
James Edward Grace
Rosa C. See

Product Manager
Phil Garcia

Publisher
Mary D. Smith, M.S. Ed.

Authors

Teacher Created Resources Staff

Teacher Created Resources, Inc.
6421 Industry Way
Westminster, CA 92683
www.teachercreated.com

ISBN-0-7439-3311-7

©2002 Teacher Created Resources, Inc.

Reprinted, 2005

Made in U.S.A.

The classroom teacher may reproduce copies of materials in this book for classroom use only. The reproduction of any part for an entire school or school system is strictly prohibited. No part of this publication may be transmitted, stored, or recorded in any form without written permission from the publisher.

Table of Contents and Index

© Teacher Created Resources, Inc.

Introduction

The old adage "practice makes perfect" can really hold true for your child and his or her education. The more practice and exposure your child has with concepts being taught in school, the more success he or she is likely to find. For many parents, knowing how to help their children can be frustrating because the resources may not be readily available. As a parent it is also difficult to know where to focus your efforts so that the extra practice your child receives at home supports what he or she is learning in school.

This book has been designed to help parents and teachers reinforce basic skills with their children. *Practice Makes Perfect* reviews basic math skills for children in grade 1. The math focus is word problems. While it would be impossible to include in this book all concepts taught in grade 1, the following basic objectives are reinforced through practice exercises. These objectives support math standards established on a district, state, or national level. (Refer to the Table of Contents for the specific objectives of each practice page.)

- adding and subtracting simple picture problems
- counting on
- adding and subtracting to 6
- adding and subtracting to 10
- adding and subtracting to 12
- adding and subtracting to 14
- adding 2-digit numbers without regrouping
- subtracting 2-digit numbers without regrouping
- solving using shapes
- solving using time
- adding and subtracting money
- finding the mystery number

There are 36 practice pages organized sequentially, so children can build their knowledge from more basic skills to higher-level math skills. To correct the practice pages in this book, use the answer key provided on pages 47 and 48. Six practice tests follow the practice pages. These provide children with multiple-choice test items to help prepare them for standardized tests administered in schools. As children complete a problem, they fill in the correct letter among the answer choices. An optional "bubble-in" answer sheet has also been provided on page 46. This answer sheet is similar to those found on standardized tests. As your child completes each test, he or she can fill in the correct bubbles on the answer sheet.

How to Make the Most of This Book

Here are some useful ideas for optimizing the practice pages in this book.

- Set aside a specific place in your home to work on the practice pages. Keep it neat and tidy with materials on hand.
- Set up a certain time of day to work on the practice pages. This will establish consistency. An alternative is to look for times in your day or week that are less hectic and more conducive to practicing skills.
- Keep all practice sessions with your child positive and constructive. If the mood becomes tense or you and your child are frustrated, set the book aside and look for another time to practice with your child. Forcing your child to perform will not help. Do not use this book as a punishment.
- Help with instructions, if necessary. If your child is having difficulty understanding what to do or how to get started, work the first problem through with him or her.
- Review the work your child has done. This serves as reinforcement and provides further practice.
- Allow your child to use whatever writing instruments he or she prefers. For example, colored pencils can add variety and pleasure to drill work.
- Pay attention to the areas in which your child has the most difficulty. Provide extra guidance and exercises in those areas. Allowing children to use drawings and manipulatives, such as coins, tiles, game markers, or flash cards, can help them grasp difficult concepts more easily.
- Look for ways to make real-life application to the skills being reinforced.

Practice 1

Solve the problems.

1.

\+ =

How many cats in all? ________

2.

\+ =

How many chickens in all?________

3.

\+ =

How many pumpkins in all?________

4.

\+ =

How many cows in all?________

5.

\+ =

How many ducks in all?________

 © Teacher Created Resources, Inc.

Practice 2

2¢	**3¢**	**4¢**	**5¢**	**6¢**

Write the subtraction sentence.

1. Dan has 8¢. He buys a [picture]. How much does he have left?

   ____ ¢ – ____ ¢ = ____ ¢

2. Pam has 5¢. She buys a [picture]. How much does she have left?

   ____ ¢ – ____ ¢ = ____ ¢

3. Jan has 6¢. She buys [picture]. How much does she have left?

   ____ ¢ – ____ ¢ = ____ ¢

4. Stella has 9¢. She buys a [picture]. How much does she have left?

   ____ ¢ – ____ ¢ = ____ ¢

5. Sam has 4¢. He buys a [picture]. How much does he have left?

   ____ ¢ – ____ ¢ = ____ ¢

6. Stan has 7¢. He buys a [picture]. How much does he have left?

   ____ ¢ – ____ ¢ = ____ ¢

7. Jim has 7¢. He buys a [picture]. How much does he have left?

   ____ ¢ – ____ ¢ = ____ ¢

8. Fran has 6¢. She buys a [picture]. How much does she have left?

   ____ ¢ – ____ ¢ = ____ ¢

Practice 3

Write a number sentence for each story. Give the answer.

1\.

Tom had 8  s.

He found 7 more.

How many  s in all?

________________  s

2\.

Jane picked up 3  s.

She found 10 more.

How many  s altogether?

________________ s

3\.

9  s

15  s

How many more  s?

________________ s

4\.

9  s

8 s

How many in all?

________________ 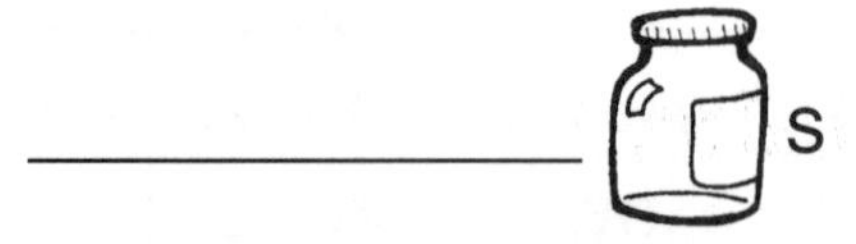 s

5\.

12  s

7  s

How many more  s?

________________  s

6\.

Mr. Tan found 16  s

He found 7 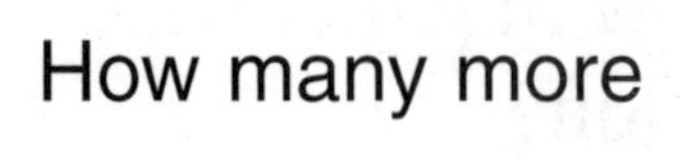 s

How many more  s?

________________ 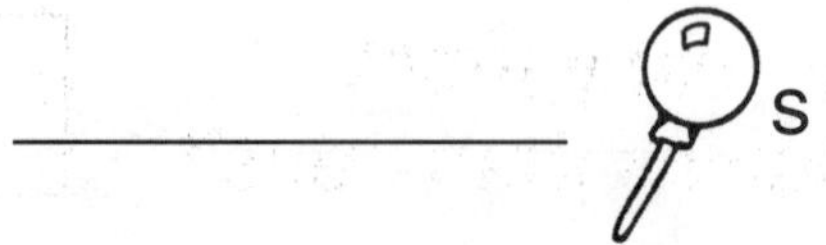 s

© Teacher Created Resources, Inc.

Practice 4

Solve the problems.

1.
4  s

3 s in each  .

How many nests in all? ☐

2.
16  s

5 more s.

How many birds in all? ☐

3.
18  s

13 s fly away.

How many owls left? ☐

4.
19  s

Only 13 have s.

How many birds do not have worms? 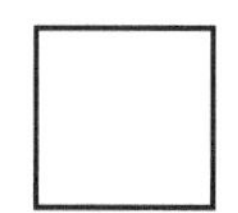

5.
37 s

12 hatch.

How many  s left? ☐

6.
28  s

3 s join them.

How many  s in all? 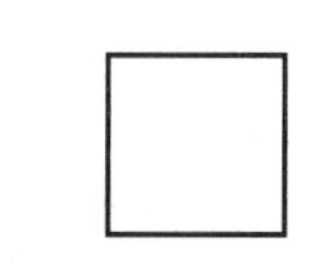

7.
11 s

6  s

How many more penguins than hens? 

8.
How many eggs are cracked?

© Teacher Created Resources, Inc.

Practice 5

Solve the problems.

1. There are 3 s on the farm. Each laid two s. How many s are there altogether?

   ____

2. Five s are growing in the field. You pick 3 of the s. How many are left?

   ____

3. One apple tree has 6 s. The other tree has 5 s. You eat one. How many apples are left? __________

4. On the farm, there are 3 s, 1 , 1 and 2 s. How many animals in all are on the farm? __________

5. The farmer wants to have 11 s in the orchard. He now has 2. How many more s will he need to plant? __________

6. Two s are in the barn. Each cow gives 3 s of milk. How many s of milk are there altogether?

   ____

© Teacher Created Resources, Inc.

Practice 6

Solve the problems.

1. Cheryl has 5 marbles. Cindy has 2 more marbles than Cheryl. How many marbles does Cindy have?

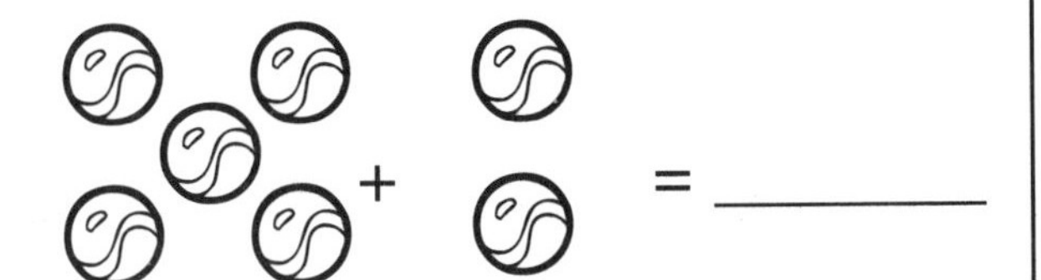

+ = ________

Cindy has ______ marbles.

2. Henry has 4 stamps. Eric has 5 more stamps than Henry. How many stamps does Eric have?

+ = ________

Eric has ______ stamps.

3. Gabby bought 1 piece of gum. Bobby bought 3 more pieces of gum than Gabby. How many pieces of gum did Bobby buy?

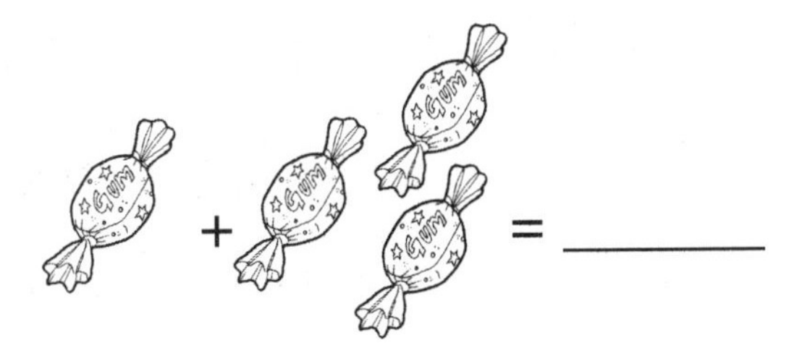

+ = ________

Bobby bought ______ pieces of gum.

4. Ana has 1 puzzle. Deanna has 5 more puzzles than Ana. How many puzzles does Deanna have?

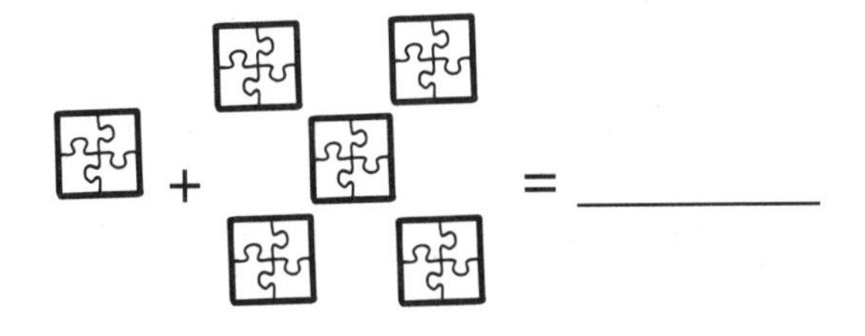

+ = ________

Deanna has ______ puzzles.

© Teacher Created Resources, Inc.

Practice 7

Solve the problems.

1. Marcy found 4 seeds in one pumpkin. She found 2 seeds in another pumpkin. How many seeds did Marcy find in all?

   + = ______

   Marcy found ______ seeds in all.

2. Mark found 1 pumpkin seed in his pocket. He found 3 pumpkin seeds in his shoe. How many pumpkin seeds did Mark find in all?

   + = ______

   Mark found ______ seeds in all.

3. Linda had 4 pumpkins. She bought 1 more. How many pumpkins does she have in all?

   Circle the answer.

   + = ______

   4 5 6 7

4. Johnny found 4 pumpkins in the hay loft. He found 2 more in the hay wagon. How many pumpkins did he find in all?

   Circle the answer.

   + = ______

   5 6 7 8

© Teacher Created Resources, Inc.

Practice 8

Solve the problems.

1. My sister bought 1 apple. Then she bought 2 more. How many apples did she buy in all?

She bought _______ apples in all.

2. My brother had 2 apple pies. Then he bought 3 more apple pies. How many apple pies did he buy in all?

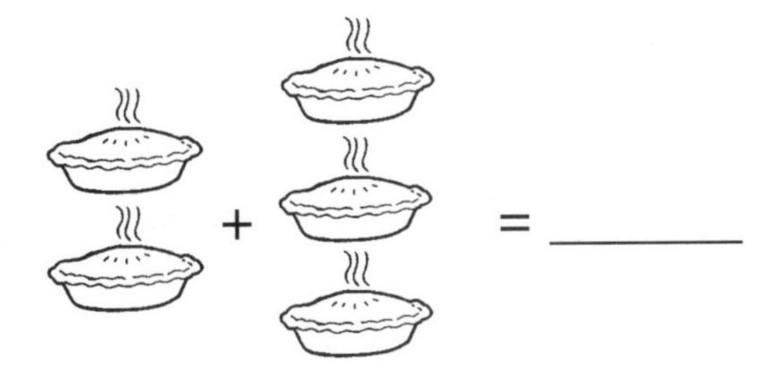

He bought _______ apple pies in all.

3. Jenna had 3 apples already. She picked 3 more off of her apple tree. How many apples does Jenna have in all?

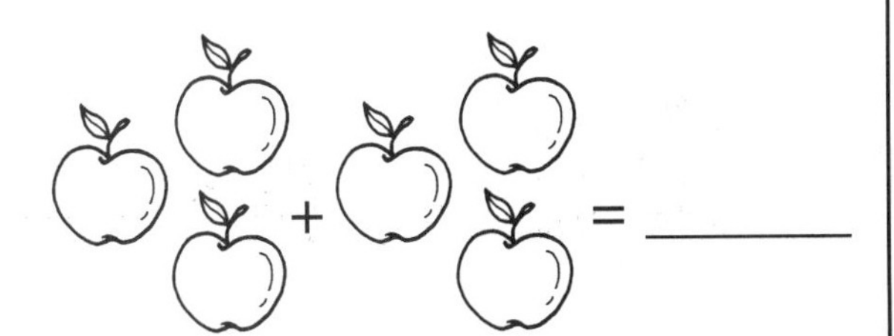

She has _______ apples in all.

4. Carmine didn't have any apples. His friend Paulo gave him 1 apple. How many apples does Carmine have now?

0 + = _______

Carmine has _______ apple now.

Practice 9

Circle *add* or *subtract*.

1. I had 4 small green apples. Jamie gave me 1 red apple. How many apples do I have in all?

add subtract

2. Gary found 1 yellow apple on the ground. He found 1 green apple on the fence. How many apples did Gary find in all?

add subtract

3. Miranda bought 3 small apples and 1 large apple. How many apples did Miranda buy in all?

add subtract

4. Barney picked 4 apples. He gave 2 of them away. How many apples does Barney have now?

add subtract

© Teacher Created Resources, Inc.

Practice 10

Solve the problems.

1. There are 5 frogs. If 1 frog hops away, how many are left?

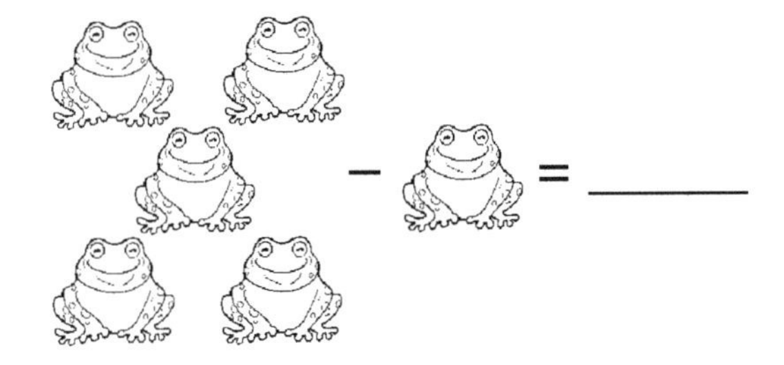

− = ______

There are ______ frogs left.

2. There are 6 frogs. If 3 frogs hop away, how many are left?

− = ______

There are ______ frogs left.

3. Jake went fishing. He caught 6 fish in the morning and 0 fish in the afternoon. How many fish did Jake catch in all?

+ 0 = ______

Jake caught ______ fish in all.

4. Ali caught 4 fish. She gave 3 to the Smith family. How many fish does Ali have left?

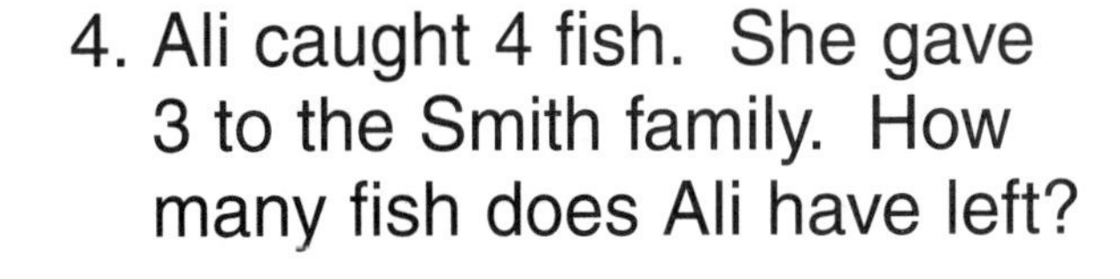

− = ______

Ali has ______ fish left.

Practice 11

Solve the problems.

1. Billy and I were outside one night. Billy saw 4 bats, and I saw 5 bats. How many bats did we see in all?

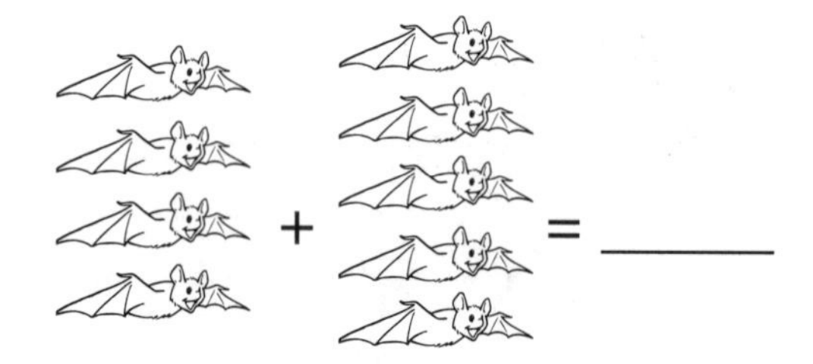

We saw ______ bats in all.

2. Susan saw 3 bats. Tasha saw 6 bats. How many bats did Susan and Tasha see in all?

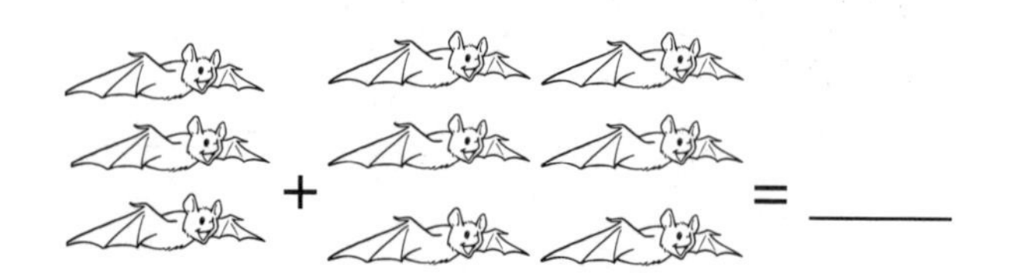

They saw ______ bats in all.

3. Aunt Clara once saw 5 fruit bats and 3 fish-eating bats. How many bats did Aunt Clara see in all?

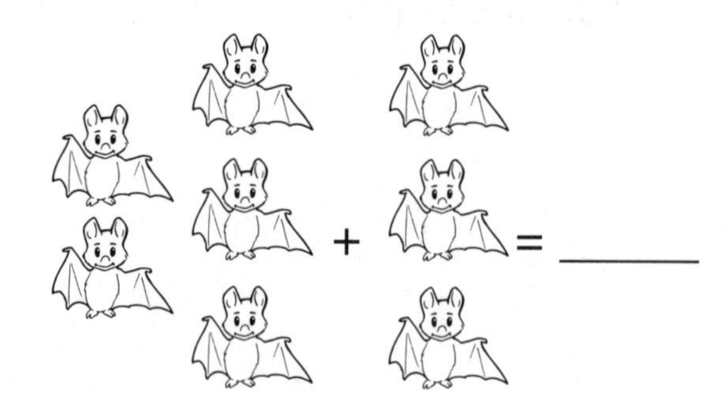

Aunt Clara saw ______ bats in all.

4. On Monday Uncle Sean saw 5 brown bats. On Tuesday he saw some more brown bats. He saw 10 bats in all. How many bats did he see on Tuesday?

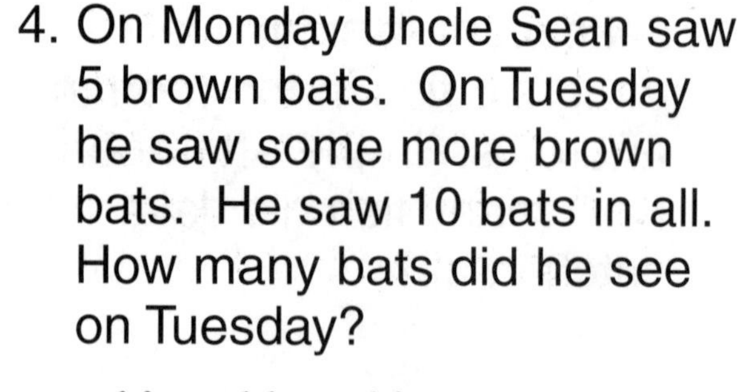

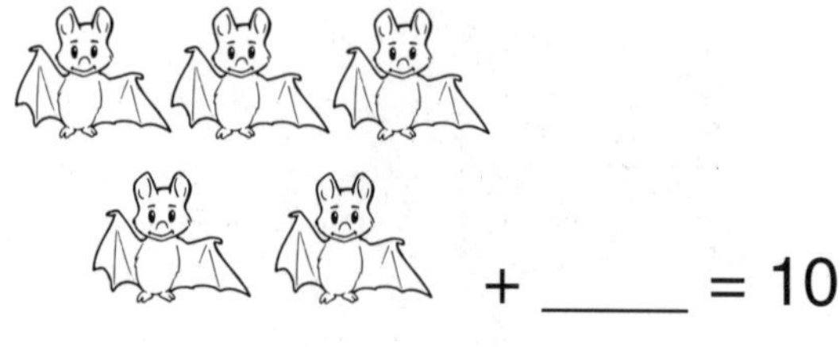

\+ _____ = 10

Monday Tuesday

Uncle Sean saw ______ bats on Tuesday.

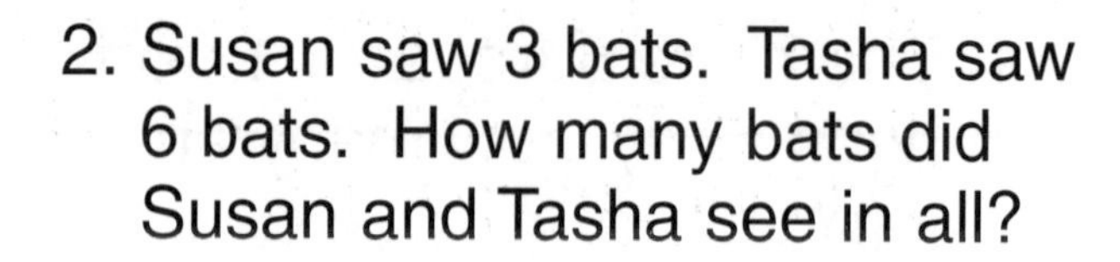

© Teacher Created Resources, Inc.

Practice 12

Solve the problems.

1. Sally put up 5 tents. Rob put up 2 tents. How many tents did they put up in all?

+ = ______

They put up ______ tents in all.

2. Zach saw 3 stars in the night sky. Rosie saw 6 stars. How many stars did they see in all?

+ = ______

They saw ______ stars in all.

3. Diego put up 2 tents in the first camp. Diego then put up 6 tents in the second camp. How many tents did Diego put up in all?

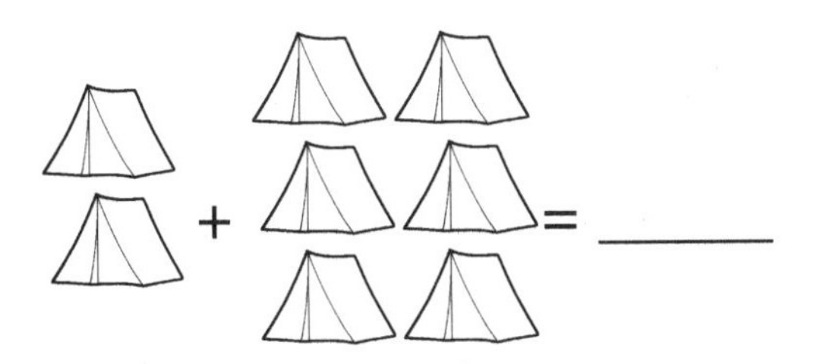

Diego put up ______ tents in all.

4. Emily found 1 rock while on her morning walk. She later found 8 rocks while on her evening walk. How many rocks did Emily find in all?

Emily found ______ rocks in all.

Practice 13

Solve the problems.

1. Ms. Grain had 8 turkeys on her farm. She sold 6 of the turkeys. How many turkeys does Ms. Grain have left?

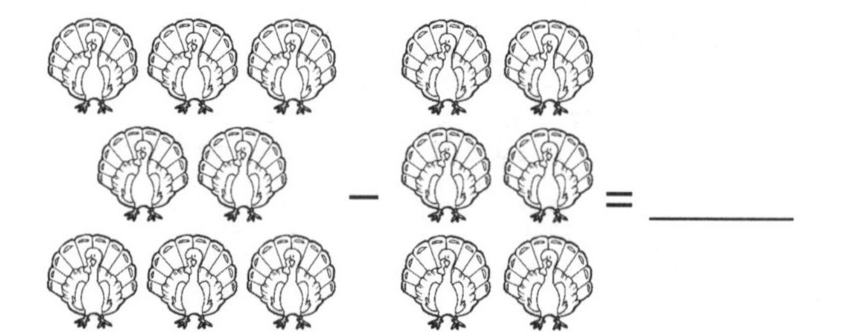

– = ______

Ms. Grain has ________ turkeys left.

2. Nick collects turkey feathers. Yesterday he collected 7 turkey feathers. He dropped 4 of them on his way home. How many feathers does Nick have left?

– = ______

Nick has ________ turkey feathers left.

3. Omar saw 7 turkeys sitting on the gate. If 3 of the turkeys flew away, how many turkeys were left?

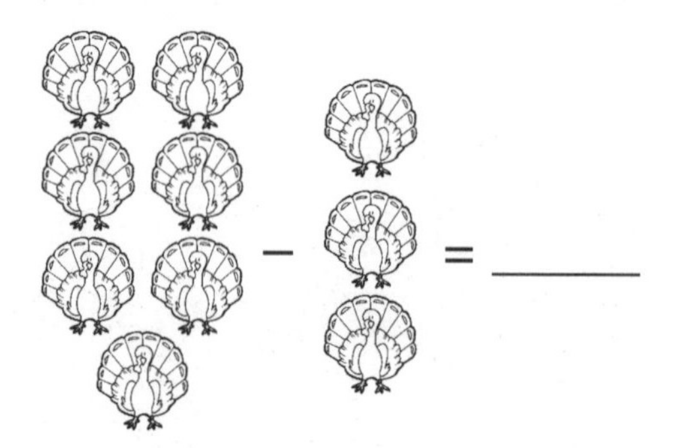

– = ______

There were ________ turkeys left.

4. Vivian gave her pet turkey 9 ears of corn. The turkey ate 2 ears of corn. How many ears of corn are left?

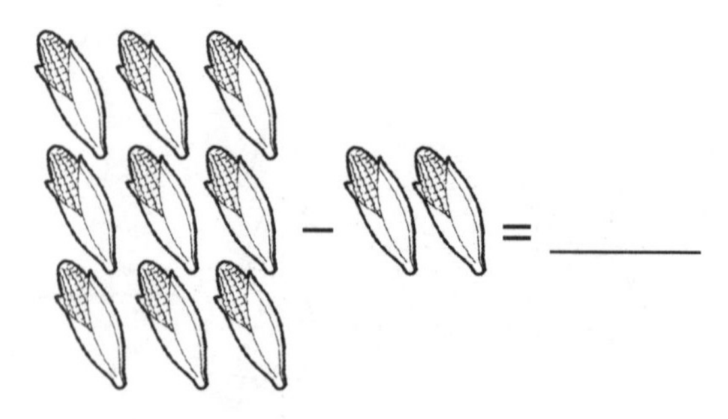

– = ______

There are ________ ears left.

© Teacher Created Resources, Inc.

Practice 14

Solve the problems.

1. There are 8 chicks. If 2 run away, how many would be left?

There would be ________ chicks left.

2. There are 10 owls. If 9 fly away, how many are left?

There would be ________ owl left.

3. Noel saw 9 spiders in the hay loft. If 3 of them were not spinning webs, how many spiders were spinning webs?

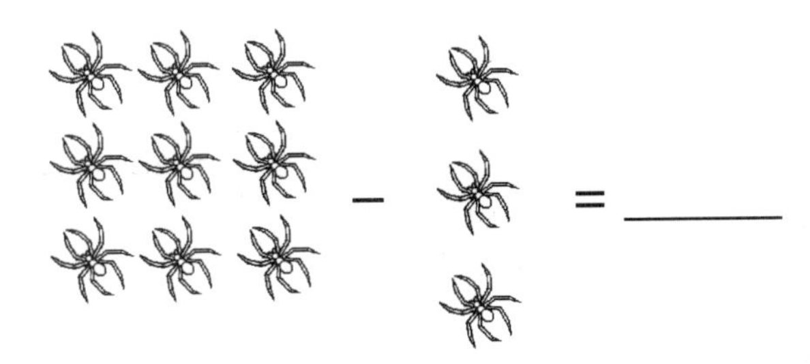

There were ________ spiders spinning webs.

4. Tim has 9 ducks. If 2 ducks waddle away, how many ducks does Tim have left?

Tim has ________ ducks left.

Practice 15

Solve the problems.

1. Amy made 1 scarecrow. Her friend gave her 8 more. How many scarecrows does Amy have in all?

   $1 + 8 =$ ______

   Amy has ______ scarecrows in all.

2. Uncle Roberto made 10 paper turkeys. He gave away 5. How many paper turkeys does Uncle Roberto have left?

   $10 - 5 =$ ______

   There are ______ paper turkeys left.

3. Mom placed 8 pumpkin pies on the windowsill. The dog ate 4 of them. How many pies are left?

   $8 - 4 =$ ______

   There are ______ pies left.

4. Sue made 9 pumpkin pies for her family. Sue's family ate 2 pumpkin pies. How many pumpkin pies does Sue have left?

   $9 - 2 =$ ______

   Sue has ______ pies left.

© Teacher Created Resources, Inc.

Practice 16

Solve the problems.

1. Ali planted 9 rows of corn on Thursday. He didn't plant any corn on Friday. How many rows of corn did Ali plant in all?

   9 + 0 = ________

   Ali planted ________ rows of corn in all.

2. Georgia had 7 ears of corn. She ate 4 ears of corn. How many ears did she have left?

   7 − 4 = ________

   She has ________ ears of corn left.

3. Enrique made 10 corn tortillas. He gave 1 corn tortilla to Herman. How many tortillas does Enrique have left?

   10 − 1 = ________

   He has ________ tortillas left.

4. Lupe had 8 corn cakes. She ate 5 of them. How many corn cakes does Lupe have left?

   8 − 5 = ________

   Lupe has ________ corn cakes left.

Practice 17

Solve the problems. Show your work.

1. Dave ate 3 doughnuts for breakfast. He ate 4 more doughnuts for snack. How many doughnuts did Dave eat in all?

   Dave ate ________ doughnuts in all.

2. Hector had 10 oranges. He ate 2. How many are left?

   ________ oranges are left.

3. Dana made 4 sandwiches. Then she made 6 more. How many sandwiches did Dana make in all?

   Dana made ________ sandwiches in all.

4. Solomon had 10 eggs. His dog ate 4 of them. How many eggs does Solomon have left?

   Solomon has ________ eggs left.

© Teacher Created Resources, Inc.

Practice 18

Solve the problems. Show your work.

1. Dareen caught 9 butterflies. Devin caught 3 butterflies. How many butterflies did they catch in all?

   They caught ________ butterflies in all.

2. Justin made 7 pictures using crayons and 4 pictures using chalk. How many pictures did Justin make in all?

   Justin made ________ pictures in all.

3. Ginger had 9 seashells. Now she has only 2. How many seashells did she lose?

   Ginger lost ________ seashells.

4. Rosemary made 11 cookies. Her brother ate some. Now there are only 2 left. How many cookies did her brother eat?

   He ate ________ cookies.

© Teacher Created Resources, Inc.

Practice 19

Solve the problems. Show your work.

1. Stuart ate 10 candies. Glenn ate 3 candies. How many candies did they eat in all?

   They ate ________ candies in all.

2. Taylor counted 8 red roses and 6 white roses in the garden. How many roses did Taylor find in all?

   Taylor found ________ roses

3. Yvonne collected 6 cans and 7 newspapers to take to the recycling center. How many items did Yvonne collect in all?

   Yvonne collected ________ items in all.

4. Lars is 7 years old. Lonnie is also 7 years old. If their ages were added together, how old would they be?

   They would be ________ years old.

© Teacher Created Resources, Inc.

Practice 20          

Solve the problems. Show your work.

1. Gwen has 10 pigs. If 3 ran into the barn, how many pigs did not run into the barn?

   ________ pigs did not run into the barn.

2. Farmer Ada had 14 animals. If 8 of them were chickens, how many were not chickens?

   ________ animals were not chickens.

3. Farmer Cary feeds the cows 9 bales of hay. He feeds the horses 4 bales of hay. How many more bales of hay does he give to the cows than to the horses?

   He gives the cows ________ more bales of hay.

4. Leo picked 12 blueberries. He gave 9 of them to his friend. How many blueberries does Leo have left?

   Leo has ________ blueberries left.

Practice 21

Solve the problems. Show your work.

1. Stacy had 12 pieces of candy. She gave 6 pieces to Raul. How many pieces of candy does Stacy have left?

   Stacy has ________ pieces left.

2. Bea has 6 candles. She buys 5 more. How many candles does Bea now have?

   Bea has ________ candles.

3. Manuel made 11 cakes. He sold 6 at the fair. How many cakes does Manuel have left?

   Manuel has ________ cakes left.

4. Omar can play 9 songs on his guitar. He has already played 3 songs. How many more songs can Omar play?

   Omar can play ________ more songs.

© Teacher Created Resources, Inc.

Practice 22

Use the hundreds chart to solve each problem.

1	2	3	4	5	6	7	8	9	10
11	12	13	14	15	16	17	18	19	20
21	22	23	24	25	26	27	28	29	30
31	32	33	34	35	36	37	38	39	40
41	42	43	44	45	46	47	48	49	50
51	52	53	54	55	56	57	58	59	60
61	62	63	64	65	66	67	68	69	70
71	72	73	74	75	76	77	78	79	80
81	82	83	84	85	86	87	88	89	90
91	92	93	94	95	96	97	98	99	100

1. I am the number that comes right after 9. What number am I?

   The mystery number is

   ________ .

2. I am the number that comes right after 5. What number am I?

   The mystery number is

   ________ .

3. I am the number that comes between 12 and 14. What number am I?

   The mystery number is

   ________ .

4. I am the number that comes between 16 and 18. What number am I?

   The mystery number is

   ________ .

Practice 23

Use the target board below to solve the problems.

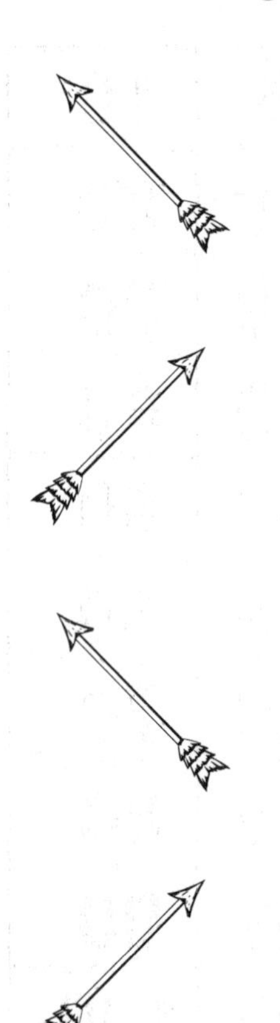

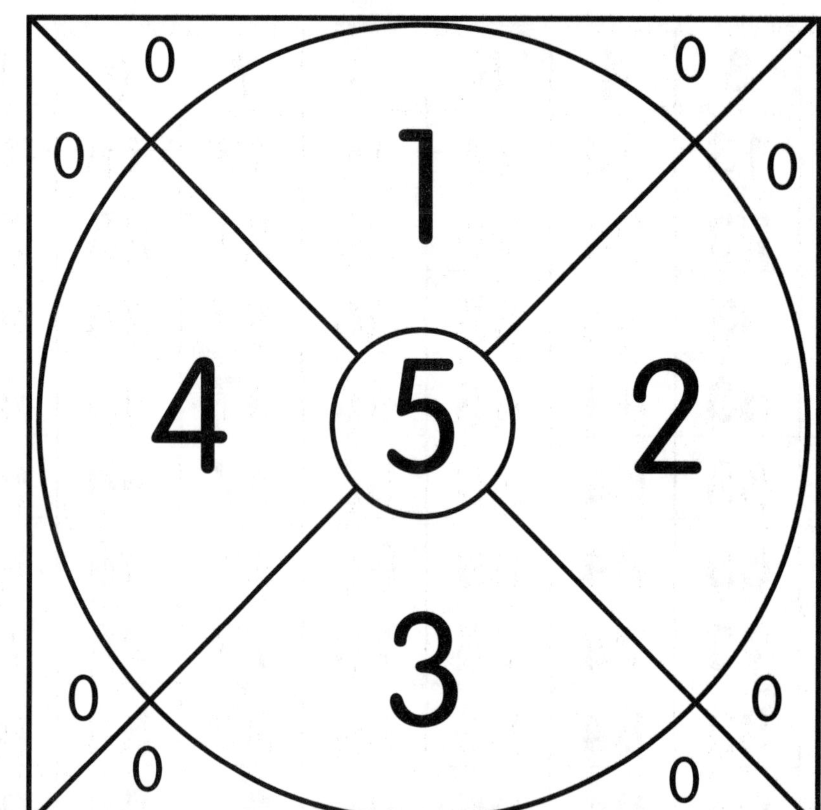

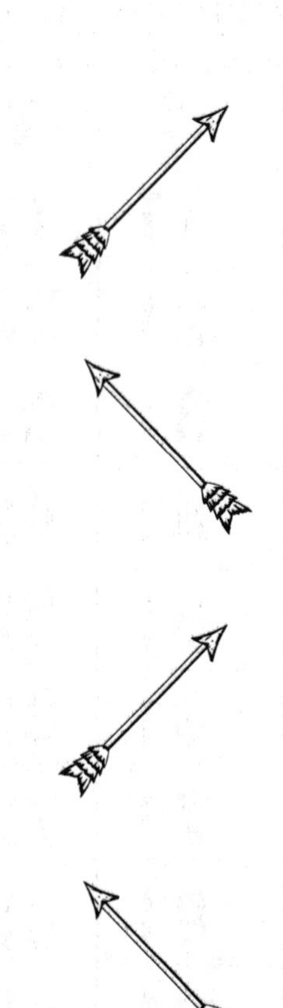

1. Robin shot two arrows for a total of 5 points. The first arrow hit the 0. What was the other number that Robin hit?

   Robin hit the ________.

2. Maid Marion shot two arrows for a total of 3 points. The first arrow hit the 2. What was the other number that Maid Marion hit?

   Maid Marion hit the ________.

3. Little John shot two arrows for a total of 10 points. The first arrow hit the 5. What was the other number that Little John hit?

   Little John hit the ________.

4. Friar Tuck shot two arrows for a total of 0 points. What were the two numbers that Friar Tuck hit?

   Friar Tuck hit the ________ and the ________.

 © Teacher Created Resources, Inc.

Practice 24

Use the hundreds chart to solve each problem.

1	2	3	4	5	6	7	8	9	10
11	12	13	14	15	16	17	18	19	20
21	22	23	24	25	26	27	28	29	30
31	32	33	34	35	36	37	38	39	40
41	42	43	44	45	46	47	48	49	50
51	52	53	54	55	56	57	58	59	60
61	62	63	64	65	66	67	68	69	70
71	72	73	74	75	76	77	78	79	80
81	82	83	84	85	86	87	88	89	90
91	92	93	94	95	96	97	98	99	100

1. I am larger than 20 and less than 40. I am an even number. When you count by tens you say my name.

   What number am I? ________

2. I am less than 80 but larger than 10. I have two numbers that are the same. My two numbers added together equal 4.

   What number am I? ________

3. I am larger than 50 and less than 100. I have a 5 in the ones place. I have a number smaller than 6 in the tens place.

   What number am I? ________

4. I have a 2 as one of my numbers. Counting by tens you say my name.

   What number am I? ________

© Teacher Created Resources, Inc.

Practice 25

Solve the problems.

1. Frank bought 10 tickets. A friend gave him 15 more. How many tickets does Frank have in all?

$$\begin{array}{r} 10 \\ +\ 15 \\ \hline \end{array}$$

Frank has ________ tickets in all.

2. Mr. Simons caught 30 fish in the morning and 40 fish in the afternoon. How many fish did he catch in all?

$$\begin{array}{r} 30 \\ +\ 40 \\ \hline \end{array}$$

Mr. Simons caught ________ fish in all.

3. Shannon made 14 baskets in the first game and 20 baskets in the second game. How many baskets did Shannon make in all?

$$\begin{array}{r} 14 \\ +\ 20 \\ \hline \end{array}$$

Shannon made ________ baskets in all.

4. Dean had a necktie with 31 red dots and 42 blue dots. How many dots in all did Dean have on his tie?

$$\begin{array}{r} 31 \\ +\ 42 \\ \hline \end{array}$$

Dean had ________ dots in all.

© Teacher Created Resources, Inc.

Practice 26

Solve the problems. Show your work.

1. Maria had 45 stamps in her collection. Her aunt gave her 54 more. How many stamps does Maria have in all?

Maria has ________ stamps in all.

2. Ferdinand planted 60 yellow tulips and 10 pink tulips. How many tulips did Ferdinand plant in all?

+ ______

Ferdinand planted ________ tulips in all.

3. Mickey has 70 baseball cards and 20 football cards. How many cards does Mickey have in all?

+ ______

Mickey has ________ cards in all.

4. My first cookie had 15 chocolate chips in it. My second cookie had 21 chocolate chips in it. How many chocolate chips were there in all?

+ ______

There were ________ chocolate chips in all.

Practice 27

Solve the problems.

1. Lisa had 73 cookies. She sold 12 of them. How many cookies does she have left?

$$\begin{array}{r} 73 \\ -\ 12 \\ \hline \end{array}$$

Lisa has ________ cookies left.

2. Tim's dog had 26 fleas. 13 of them jumped off. How many fleas are left on Tim's dog?

$$\begin{array}{r} 26 \\ -\ 13 \\ \hline \end{array}$$

Tim's dog has ________ fleas left.

3. Buster looked in his toy box and found 87 marbles. He gave 10 to Tom. How many marbles does Buster have left?

$$\begin{array}{r} 87 \\ -\ 10 \\ \hline \end{array}$$

Buster has _______ marbles left.

4. Mark had 35 pairs of white socks. He lost 13 pairs of white socks on vacation. How many pairs of white socks does Mark have left?

$$\begin{array}{r} 35 \\ -\ 13 \\ \hline \end{array}$$

Mark has ________ pairs of white socks left.

© Teacher Created Resources, Inc.

Practice 28

Solve the problems. Show your work.

1. Dad had 23 neckties. He gave away 11 ties. How many ties does Dad have now?

   –
   ______

   Dad now has ________ ties.

2. Pao had 19 tires. He sold 15 of them. How many tires does Pao have left?

   –
   ______

   Pao has ________ tires left.

3. Sara had 17 cream pies. She sells 11 of them. How many pies does Sara have now?

   –
   ______

   Sara now has ________ pies.

4. Louisa bought a dozen eggs. If 6 of the eggs broke, how many eggs does Louisa have left?

   –
   ______

   Louisa has ________ eggs left.

© Teacher Created Resources, Inc.

Practice 29

Circle *add* or *subtract*.

1. Polly counted 49 insects. If 36 of them flew away, how many are left?

   add

   subtract

2. William had 71 red spots on one arm and 28 red spots on the other arm. How many spots did he have in all?

   add

   subtract

3. Kim picked 27 flowers. If 16 wilted on the way home, how many are still fresh?

   add

   subtract

4. Tracy had 38 letters. She mailed 24 of them. How many are left?

   add

   subtract

© Teacher Created Resources, Inc.

Practice 30

Use the shapes to solve the problems.

triangle **circle** **square**

rectangle **hexagon** **pentagon**

1. I have 3 sides and 3 corners. Which shape am I?

   ____

2. I have 5 sides and 5 corners. Which shape am I?

   ____

3. I have 6 sides and 6 corners. Which shape am I?

   ____

4. I have 2 long sides, 2 short sides, and 4 corners. Which shape am I?

   ____

5. I do not have any corners. Which shape am I?

   ____

6. I have 4 corners and 4 equal sides. Which shape am I?

   ____

© Teacher Created Resources, Inc.

Practice 31

1. Raul takes a 1-hour nap each day. He fell asleep at 1:00. What time does Raul wake up?

   Show the time on the clock.

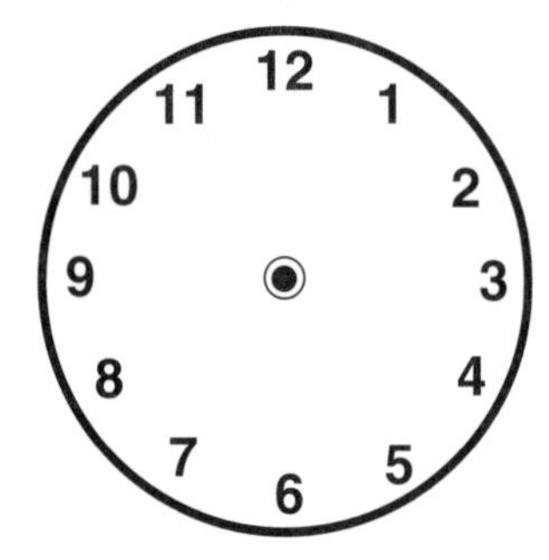

2. Fran went to the 3:00 movie. The movie was 2 hours long. What time did the movie end?

   Show the time on the clock.

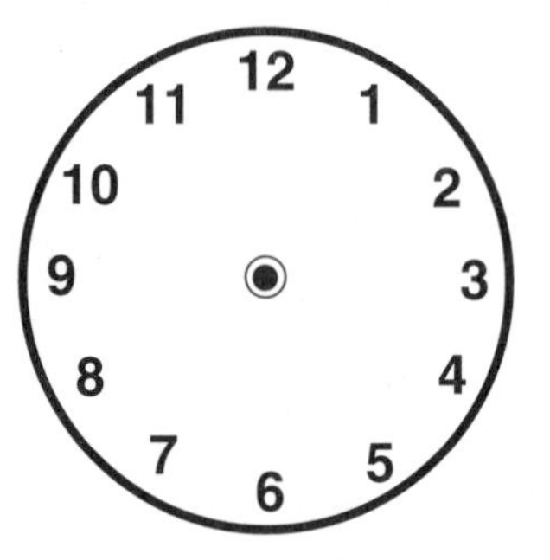

3. Phil eats dinner at 8:00 P.M. and goes to bed 1/2 hour later. What time does Phil go to bed?

   Show the time on the clock.

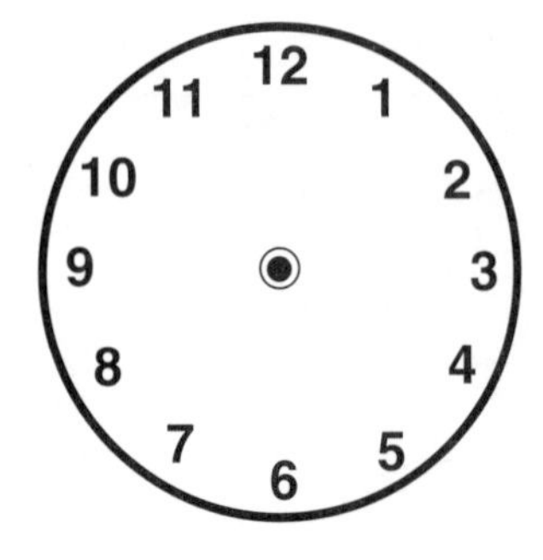

4. Marianne gets up at 7:00 A.M. and goes to school 1/2 hour after she gets up. What time does she go to school?

   Show the time on the clock.

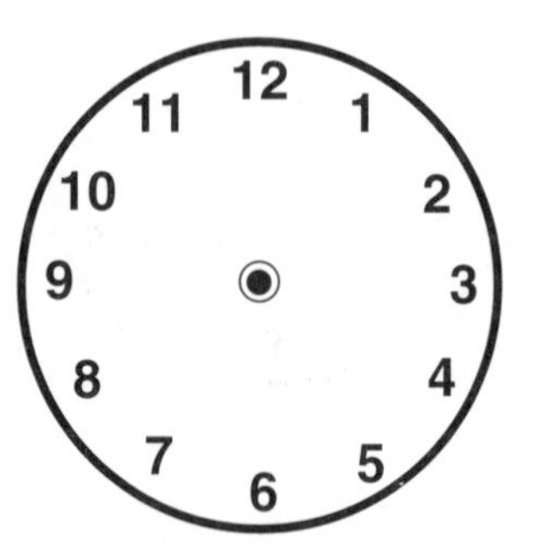

 © Teacher Created Resources, Inc.

Practice 32

1. The play began at 6:00 and lasted 1/2 hour. What time did the play end?

   Show the time and write it.

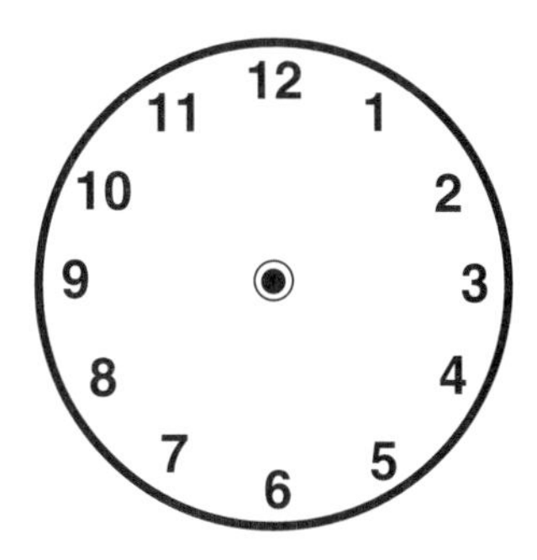

________ : ________

2. The school bus picked us up at 2:00. The bus ride took 1/2 hour. What time did we get home?

   Show the time and write it.

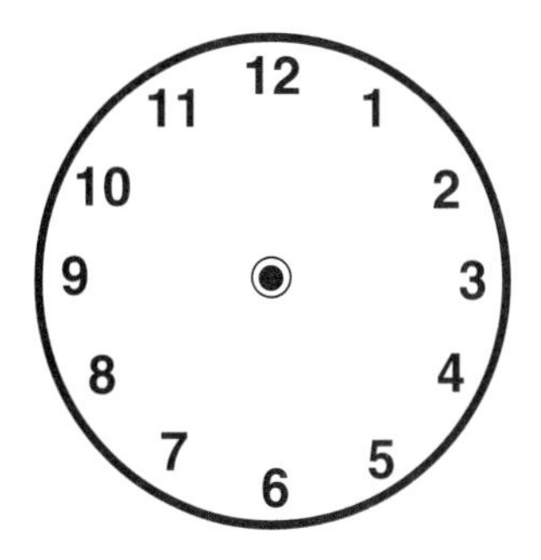

________ : ________

3. We went to the park at 4:00. We played for 1/2 hour. What time did we leave the park?

   Show the time and write it.

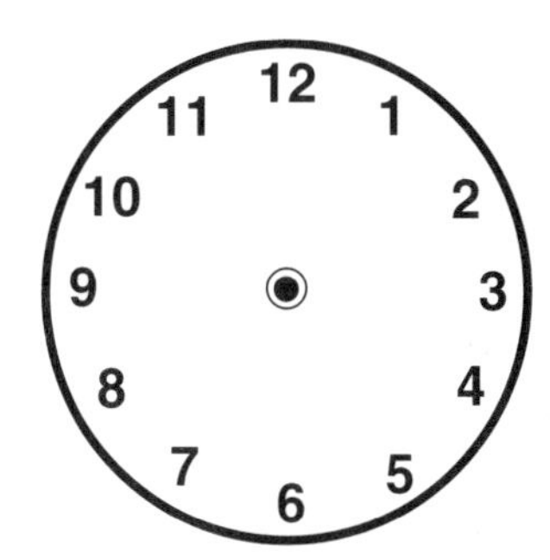

________ : ________

4. The baby fell asleep at 9:00 and napped for 1/2 hour. What time did the baby wake up?

   Show the time and write it.

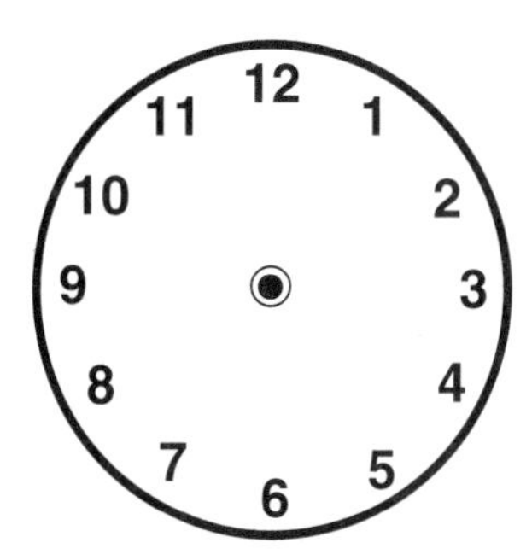

________ : ________

© Teacher Created Resources, Inc.

Practice 33

Use the money chart to help you solve the problems.

1¢	**5¢**	**10¢**	**25¢**

1. Jerome has 1 nickel in his pocket. How much money does Jerome have?

   Circle the answer.

   5¢ 10¢ 25¢

2. C.J. has 1 quarter in her pocket. How much money does C.J. have?

   Circle the answer.

   5¢ 10¢ 25¢

3. Mabel has two pockets. In one pocket she has 1 nickel. In the other pocket she has 1 dime. How much money does Mabel have?

   Circle the answer.

   5¢ 10¢ 15¢

4. Neil has two pockets, too. In one pocket he has 1 nickel. In the other pocket he has 1 nickel also. How much money does Neil have?

   Circle the answer.

   5¢ 10¢ 15¢

 © Teacher Created Resources, Inc.

Practice 34

Use the money chart to help you solve the problems.

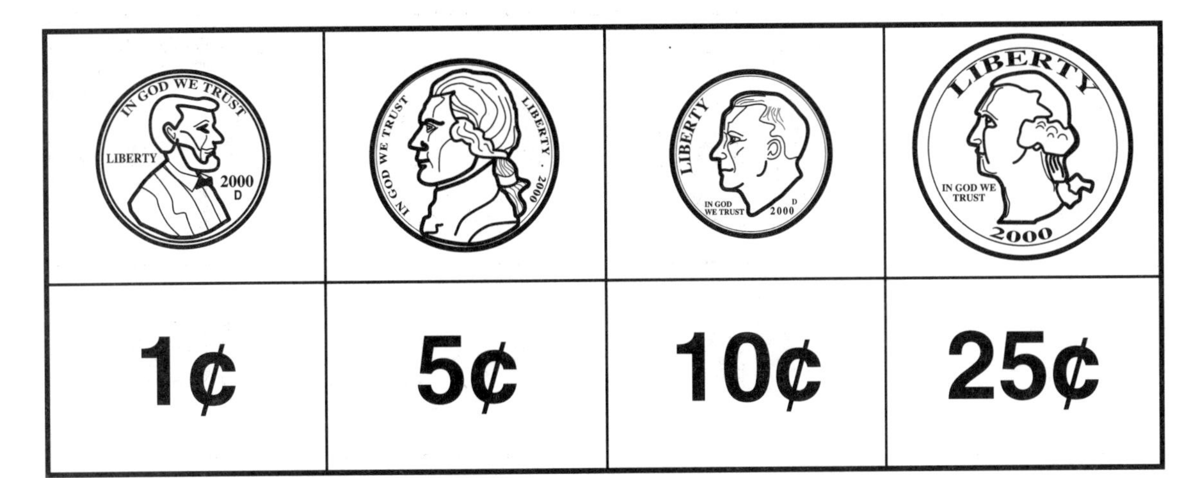

1. Xavier found 6¢ in his shirt pocket and 3¢ in his jacket pocket. How much money did Xavier find in all?

   6¢ + 3¢ = _________¢

   Xavier found _________¢ in all.

2. Maria had 5¢ in her piggy bank and 3¢ in her purse. How much money did Maria have in all?

   5¢ + 3¢ = _________¢

   Maria has _________¢ in all.

3. I have 2 coins that make exactly 6¢. One of the coins is a nickel. What is the other coin?

   The other coin is a

   ___________________________.

4. I have 2 coins that make exactly 10¢. Both of the coins are the same. What are the two coins that I have?

   The two coins are both

   ___________________________.

© Teacher Created Resources, Inc.

Practice 35

Use the money chart to help you solve the problems.

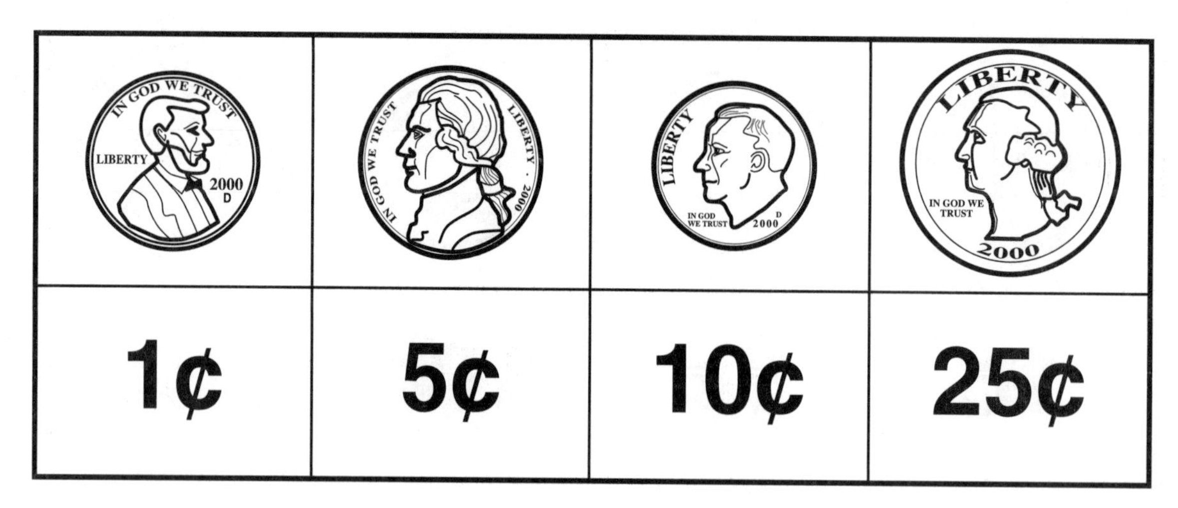

1. Stephanie had 18¢. She spent 10¢ on an ice cream cone. How much money does Stephanie have left?

   18¢
   – 10¢

   Stephanie has ________¢ left.

2. Tyrone has 36¢. He bought a model car for 25¢. How much money does Tyrone have left?

   36¢
   – 25¢

   Tyrone has ________¢ left.

3. Imani had a dime and 2 nickels. She bought candy that cost 10¢. How much money did she have left?

   20¢
   – 10¢

   Imani had ________¢ left.

4. Esperanza had 2 quarters and 3 pennies. She bought a toy for 21¢. How much money did she have left?

   ¢
   – ¢

   Esperanza had ________¢ left.

Practice 36

Use the money chart to help you solve the problems.

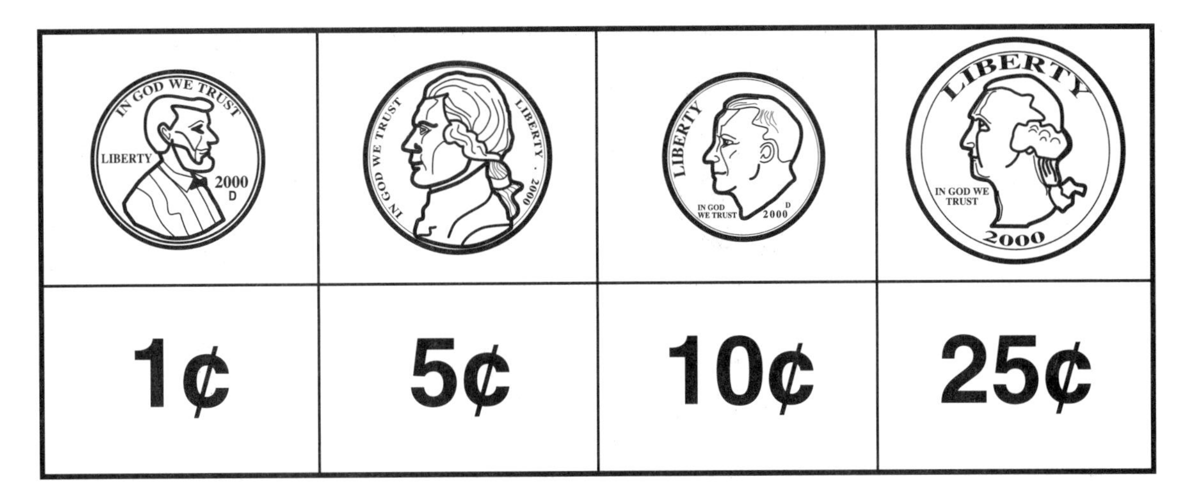

1. Ashley has a quarter and a nickel in her pocket. How much money does she have in all?

   \+ ______

   Ashley has ________¢ in all.

2. Vincent had 35¢. He spent 10¢ buying a comic book. How much money does Vincent have left?

   − ______

   Vincent has ________¢ left.

3. Wesley borrowed 8¢ from his mom and 10¢ from his dad. How much money did Wesley borrow in all?

   \+ ______

   Wesley borrowed ________¢ in all.

4. Veronica had 18¢. She lent one friend a nickel. How much money does Veronica have left?

   − ______

   Veronica has ________¢ left.

Test Practice 1

Fill in the correct answer bubble.

1. Dale made 2 red hearts. Then he made 2 more. How many hearts did Dale make in all?

   Ⓐ 0 Ⓑ 4

   Ⓒ 1 Ⓓ 3

2. Dale made 5 apple pies. He sold 1 pie. How many pies does Dale have left?

   Ⓐ 4 Ⓑ 6

   Ⓒ 5 Ⓓ 3

3. Meg has 2 bows. She buys 4 more. How many does she now have?

   Ⓐ 2 Ⓑ 3

   Ⓒ 5 Ⓓ 6

4. Leanne has 9 birds. If 3 birds fly away, how many birds are left?

   Ⓐ 0 Ⓑ 6

   Ⓒ 3 Ⓓ 5

 © Teacher Created Resources, Inc.

Test Practice 2 

Fill in the correct answer bubble.

1. There were 9 children playing outside. If 8 children went inside, how many children were still playing?

   Ⓐ 0 Ⓑ 4

   Ⓒ 1 Ⓓ 3

2. Saed has 5 books. He checked out 3 more books from the library. How many books does Saed have in all?

   Ⓐ 0 Ⓑ 2

   Ⓒ 1 Ⓓ 8

3. My teacher fixed 3 clocks yesterday and 7 clocks today. How many clocks did my teacher fix in all?

   Ⓐ 5 Ⓑ 4

   Ⓒ 10 Ⓓ 3

4. Larry stacked up 8 books. If 4 of the books fell over, how many books are left in the stack?

   Ⓐ 4 Ⓑ 3

   Ⓒ 1 Ⓓ 8

Test Practice 3

Fill in the correct answer bubble.

1. The children made 55 blueberry pancakes and 41 buttermilk pancakes. How many pancakes did they make in all?

   Ⓐ 14 Ⓑ 86

   Ⓒ 13 Ⓓ 96

2. There are 99 days in winter. If 15 days have passed, how many more days of winter are left?

   Ⓐ 84 Ⓑ 114

   Ⓒ 74 Ⓓ 48

3. My cousin has 63 coins in his piggy bank. I have 21 coins in my piggy bank. How many coins do we have in all?

   Ⓐ 84 Ⓑ 86

   Ⓒ 42 Ⓓ 48

4. At the pie-eating contest, there were 67 pies. Leslie's team ate 31 pies. How many pies were left?

   Ⓐ 84 Ⓑ 9

   Ⓒ 26 Ⓓ 36

© Teacher Created Resources, Inc.

Test Practice 4

Fill in the correct answer bubble.

1. I have 3 sides and 3 corners. Which shape am I?

Ⓐ square Ⓑ triangle

Ⓒ circle Ⓓ rectangle

2. I have 6 sides and 6 corners. Which shape am I?

Ⓐ rectangle Ⓑ hexagon

Ⓒ pentagon Ⓓ octagon

3. Mary went to bed at 2:00 and napped for 1 hour. What time did Mary wake up? Use the clock to help you choose.

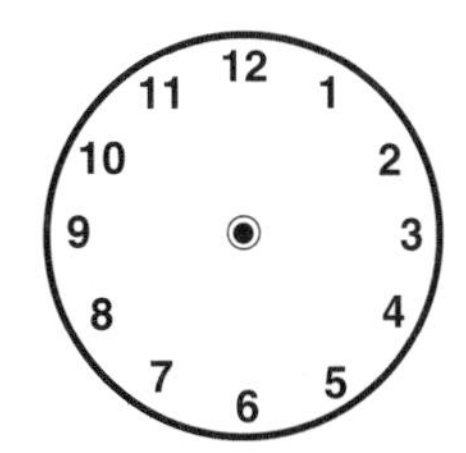

Ⓐ 2:30 Ⓑ 3:00

Ⓒ 1:30 Ⓓ 2:45

4. James eats dinner at 6:00 and finishes in 1/2 hour. What time does James finish his dinner? Use the clock to help you choose.

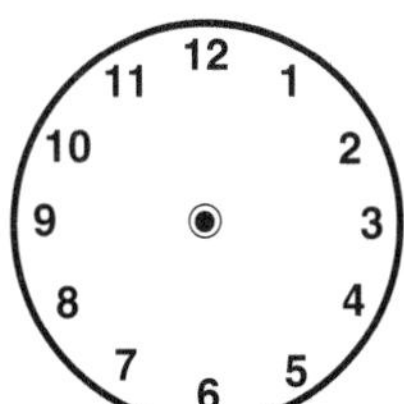

Ⓐ 6:30 Ⓑ 6:00

Ⓒ 5:30 Ⓓ 6:45

Test Practice 5

Fill in the correct answer bubble.

1. Herb had 26¢. He spent 13¢ buying a candy bar. How much money does Herb have left?

   Ⓐ 39¢ Ⓑ 12¢

   Ⓒ 19¢ Ⓓ 13¢

2. Betty had 45¢ in one pocket and 33¢ in another pocket. How much money does Betty have?

   Ⓐ 12¢ Ⓑ 78¢

   Ⓒ 68¢ Ⓓ 11¢

3. Dale had 57¢. He found 12¢ in his pocket. How much money does Dale have?

   Ⓐ 55¢ Ⓑ 69¢

   Ⓒ 45¢ Ⓓ 49¢

4. Fay had 89¢. She spent 70¢ buying a CD. How much money does Fay have left?

   Ⓐ 15¢ Ⓑ 18¢

   Ⓒ 19¢ Ⓓ 20¢

© Teacher Created Resources, Inc.

Test Practice 6

Fill in the correct answer bubble.

1. I am greater than 10 and less than 14. I am an even number. What number am I?

   Ⓐ 14 Ⓑ 10

   Ⓒ 11 Ⓓ 12

2. I am greater than 15 and less than 25. When you count by tens you say my name. What number am I?

   Ⓐ 20 Ⓑ 4

   Ⓒ 1 Ⓓ 30

3. I am an even number. I am greater than 30 and less than 50. When you count by tens you say my name. What number am I?

   Ⓐ 35 Ⓑ 40

   Ⓒ 45 Ⓓ 50

4. I am an odd number. I am greater than 80 and less than 90. When you count by fives you say my name. What number am I?

   Ⓐ 80 Ⓑ 85

   Ⓒ 90 Ⓓ 87

Answer Sheet

Test Practice 1 (Page 40)	Test Practice 2 (Page 41)	Test Practice 3 (Page 42)
1. Ⓐ Ⓑ Ⓒ Ⓓ	1. Ⓐ Ⓑ Ⓒ Ⓓ	1. Ⓐ Ⓑ Ⓒ Ⓓ
2. Ⓐ Ⓑ Ⓒ Ⓓ	2. Ⓐ Ⓑ Ⓒ Ⓓ	2. Ⓐ Ⓑ Ⓒ Ⓓ
3. Ⓐ Ⓑ Ⓒ Ⓓ	3. Ⓐ Ⓑ Ⓒ Ⓓ	3. Ⓐ Ⓑ Ⓒ Ⓓ
4. Ⓐ Ⓑ Ⓒ Ⓓ	4. Ⓐ Ⓑ Ⓒ Ⓓ	4. Ⓐ Ⓑ Ⓒ Ⓓ

Test Practice 4 (Page 43)	Test Practice 5 (Page 44)	Test Practice 6 (Page 45)
1. Ⓐ Ⓑ Ⓒ Ⓓ	1. Ⓐ Ⓑ Ⓒ Ⓓ	1. Ⓐ Ⓑ Ⓒ Ⓓ
2. Ⓐ Ⓑ Ⓒ Ⓓ	2. Ⓐ Ⓑ Ⓒ Ⓓ	2. Ⓐ Ⓑ Ⓒ Ⓓ
3. Ⓐ Ⓑ Ⓒ Ⓓ	3. Ⓐ Ⓑ Ⓒ Ⓓ	3. Ⓐ Ⓑ Ⓒ Ⓓ
4. Ⓐ Ⓑ Ⓒ Ⓓ	4. Ⓐ Ⓑ Ⓒ Ⓓ	4. Ⓐ Ⓑ Ⓒ Ⓓ

© Teacher Created Resources, Inc.

Answer Key

Page 4
1. 3
2. 4
3. 5
4. 2
5. 6

Page 5
1. 8¢ – 4¢ = 4¢
2. 5¢ – 5¢ = 0¢
3. 6¢ – 3¢ = 3¢
4. 9¢ – 5¢ = 4¢
5. 4¢ – 2¢ = 2¢
6. 7¢ – 6¢ = 1¢
7. 7¢ – 4¢ = 3¢
8. 6¢ – 6¢ = 0¢

Page 6
1. 8 + 7 = 15
2. 3 + 10 = 13
3. 15 – 9 = 6
4. 9 + 8 = 17
5. 12 – 7 = 5
6. 16 – 7 = 9

Page 7
1. 12
2. 21
3. 5
4. 6
5. 25
6. 31
7. 5
8. 4

Page 8
1. 6
2. 2
3. 10
4. 7
5. 9
6. 6

Page 9
1. 7
2. 9
3. 4
4. 6

Page 10
1. 6
2. 4
3. 5
4. 6

Page 11
1. 3
2. 5
3. 6
4. 1

Page 12
1. add
2. add
3. add
4. subtract

Page 13
1. 4
2. 3
3. 6
4. 1

Page 14
1. 9
2. 9
3. 8
4. 5

Page 15
1. 7
2. 9
3. 8
4. 9

Page 16
1. 2
2. 3
3. 4
4. 7

Page 17
1. 6
2. 1
3. 6
4. 7

Page 18
1. 9
2. 5
3. 4
4. 7

Page 19
1. 9
2. 3
3. 9
4. 3

Page 20
1. 7
2. 8
3. 10
4. 6

Page 21
1. 12
2. 11
3. 7
4. 9

Page 22
1. 13
2. 14
3. 13
4. 14

Page 23
1. 7
2. 6
3. 5
4. 3

Page 24
1. 6
2. 11
3. 5
4. 6

Page 25
1. 10
2. 6
3. 13
4. 17

Page 26
1. 5
2. 1
3. 5
4. 0, 0

Page 27
1. 30
2. 22
3. 55
4. 20

Page 28
1. 25
2. 70
3. 34
4. 73

Page 29
1. 99
2. 70
3. 90
4. 36

Page 30
1. 61
2. 13
3. 77
4. 22

Page 31
1. 12
2. 4
3. 6
4. 6

Page 32
1. subtract
2. add
3. subtract
4. subtract

Page 33
1. triangle
2. pentagon
3. hexagon
4. rectangle
5. circle
6. square

Answer Key

Page 34

1.

2.

3.
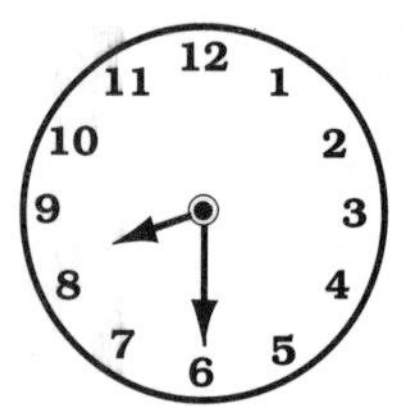

4.

Page 35

1.

6:30

2.
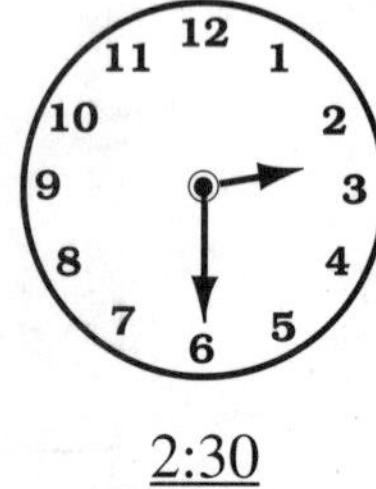

2:30

Page 35 *(cont.)*

3.

4:30

4.
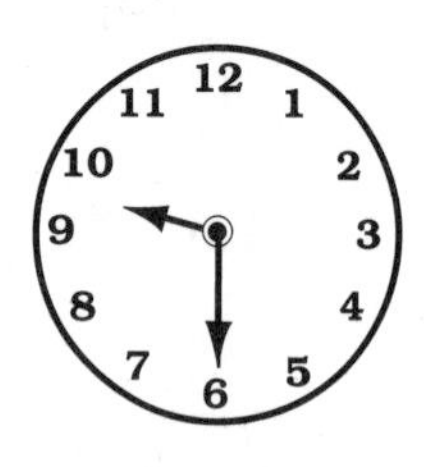

9:30

Page 36

1. 5¢
2. 25¢
3. 15¢
4. 10¢

Page 37

1. 9¢
2. 8¢
3. penny
4. nickels

Page 38

1. 8¢
2. 11¢
3. 10¢
4. 32¢

Page 39

1. 30¢
2. 25¢
3. 18¢
4. 13¢

Page 40

1. (A) ● (C) (D)
2. ● (B) (C) (D)
3. (A) (B) (C) ●
4. (A) ● (C) (D)

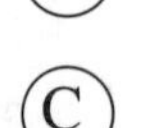
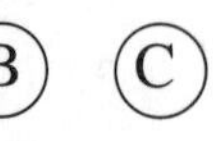

Page 41

1. (A) (B) ● (D)
2. (A) (B) (C) ●
3. (A) (B) ● (D)
4. ● (B) (C) (D)

Page 42

1. (A) (B) (C) ●
2. ● (B) (C) (D)
3. ● (B) (C) (D)
4. (A) (B) (C) ●

Page 43

1. (A) ● (C) (D)
2. (A) ● (C) (D)
3. (A) ● (C) (D)
4. ● (B) (C) (D)

Page 44

1. (A) (B) (C) ●
2. (A) ● (C) (D)
3. (A) ● (C) (D)
4. (A) (B) ● (D)

Page 45

1. (A) (B) (C) ●
2. ● (B) (C) (D)
3. (A) ● (C) (D)
4. (A) ● (C) (D)

© Teacher Created Resources, Inc.